Great Steam Trains

Mallard and the A4-Pacifics

The Loco that Wanted to Be a Racing Car

Tom Farris

Hamilton-Vale Publishing

© 2022 Tom Farris

ISBN 978-1-84285-560-7

Published by Hamilton-Vale Publishing,
New Lamorna House, Abergele LL22 7DY
www.graham-lawler.com

All rights reserved. No part of this book may be reproduced or stored in an information retrieval system without the express permission of the publishers given in writing.

The moral rights of the author have been asserted.

In the production of this book we have sourced images, some of which we understand to be copyright-free / public-domain images. This matter is then dealt with under the fair use / fair deal sections 29/30 Copyright, Designs and Patent Act 1988. In the event of a copyright claim, claimants are invited to contact the publisher, with appropriate evidence, and we will happily amend further editions.

Designed and typeset in Wales, UK
Printed in Europe.

This book is produced for educational and entertainment purposes. Readers are advised that they should not enter into any binding agreements on the basis of material in this book, without taking appropriate professional advice. Neither the author nor publisher nor any/all of the publisher's agents can be held responsible for any subsequent outcomes.

Mallard

That's a type of duck, isn't it?

Male and female Mallard ducks. The pretty one is the male. ("Mallard Ducks" by Manjith Kainickara *manjithkaini.net*)

So what has a Mallard duck to do with one of the most famous British locomotives ever made? Then what has the vast Pacific Ocean on the west side of America got to do with it?

It takes a story of Britain learning from the USA, Germany, Finland, France and other countries, then building on what they'd done for more speed – and winning – to find out.

Flashing By

Who doesn't want to get where they are going as fast they can? With the world flashing by, you get glimpses of other people's lives. It's exciting. You feel good. You want to keep going.

If you want to go fast you're best going a long way, otherwise the journey will be over before too soon. So how does nearly 640 kilometres (almost 400 miles) sound? At 160 km/h (100 mph) that would give over 4 hours of fun. The best way to do it is by train. Where in Britain could a railway find that length of journey?

It had actually been found in the 1830s between two capital cities. London in England, and Edinburgh in Scotland, The next question was how could a railway firm build such a long track?

On the east side of England it wasn't done by one firm, but by three small ones. They connected end-to-end. This became the *East Coast Main Line*. After working like this for nearly a hundred years, in 1923 these three railways became the *London and North Eastern Railway* (LNER). That's when railway travel started to get exciting, but then something else did too.

Trains v. Cars

By the 1930s, more and more people had started buying cars and were chugging around the country in them. Aeroplanes were carrying people not just abroad but quickly up and down the country too. The LNER were afraid people would be using cars and aeroplanes instead of using their trains. If the LNER didn't do something about it, they would be running empty trains, then no trains at all. How to get people out of little metal boxes on wheels or tubes in the sky and get back to filling trains?

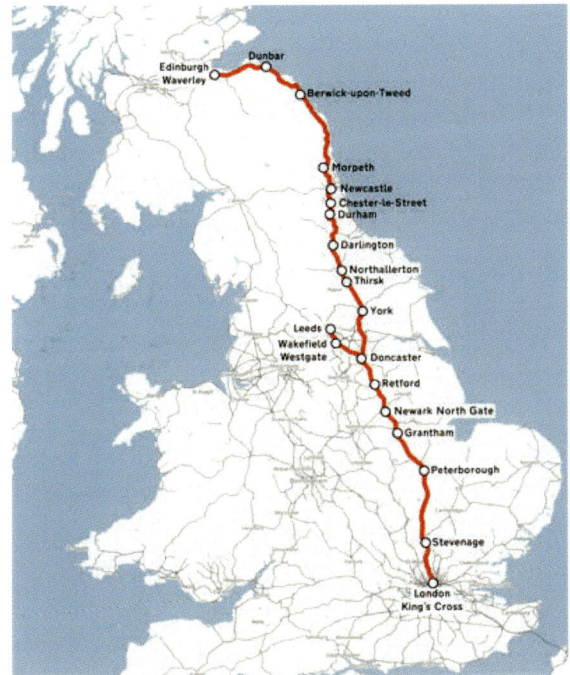

East Coast Main Line

They had some good points. Cars that most people could afford weren't as comfy, quiet or easy to drive as the ones we have today. They were also much slower. People drove at around 60 km/h (40 mph) on the open road. There were no motorways, or bypasses. They had to crawl through many towns and villages on the way at under 50 km/h (30 mph). How long to drive between London and Edinburgh, then?

It would have taken around 20 hours. It would be London people eager to visit Scotland, or Edinburgh people wanting to get home as quickly as they could. The journey could be longer if they'd got held up in traffic jams.

An Austin-6 family car of 1934.
(Edited photo by Adrian Pingstone. Public Domain.)

Driver and passengers would get out rubbing stiff backs and knees from the cramped space. In the winter the windows would get steamed up, and the heater wouldn't reach the rear seats. Bored children in the back with cold feet and hardly any view would have been squabbling, and asking, "Are we there yet?" or "How much longer?" Then there's the cry that parents dreaded, "Mummy, I need the toilet!" Could trains do better?

The LNER thought so but they needed something extra to tempt people back onto their trains. How about travelling many times faster than cars? How about no traffic jams? How about big comfy seats, toilets in each carriage, and a restaurant? Space to stretch your legs and for children to run around and annoy other passengers. Warm feet. Big windows to look out of. That adds up to luxury that people couldn't get in their cars or on aeroplanes for that matter. The trouble was that people saw the grubby, smelly, outside of trains first. A big clean-up was needed to pull people in.

How about a clean, sleek, streamlined train that oozes speed and luxury? Great idea, but who was to design it?

The Winning Teams

When the LNER was set up, they had to decide which people from the small railways to keep. One of these people was to find himself in a fantastic new job. This job was *Chief Mechanical Engineer* (CME) of the whole railway. The LNER had a choice of several great engineers. Who was it to be? There was one engineer who stood out.

His name was Nigel Gresley. He'd been the CME of the *Great Northern Railway* (GNR). This was one of those small railways. He was chosen because of the really good work he'd been doing. This was building and leading six great teams of workshop engineers and design engineers (also called *Draughtsmen*). They had already designed and built some great locomotives. It made sense to keep these teams with this new job.

Now Gresley was in the LNER, his office was in London, nearly 170 miles away from his six teams of engineers in Doncaster. He was one of the bosses, wearing a posh suit, sitting in a nice office and having lots of worries. He didn't actually get his hands dirty any more. The engineers did that.

Each of his teams had a top-notch leader Gresley could rely on. These leaders relied on their office ladies to help them manage the 30 or so hard-working people there in Doncaster. Meanwhile in London, Gresley was helped by a small team of advisors. Even though Gresley and his teams were so far apart. they had designed many great locomotives.

Who was Nigel Gresley?

Gresley had come from a posh family in Derbyshire. It was unusual for someone like that to go into engineering. After school and college, he became an apprentice (which means *learner*) at the Crewe Works of the *London and North Western Railway* (LNWR) — some railway companies designed and built their own locomotives. There he would have had lessons, as well as help from engineers. After a while, Gresley took a step up by moving to another railway.

Sir Nigel Gresley. (Photo LNER Encyclopedia, unlicensed, fair use.)

This was called the *Lancashire and Yorkshire Railway* (L&YR). There he worked under a *Chief Mechanical Engineer* not many people know of. His name was John Aspinall, who became Sir John Aspinall. He had led a team of engineers designing locomotives, as well as coming up with a way to make trains stop quicker. This was good thinking! When he'd learnt all he could from Aspinall, Gresley took another step up.

He went to the *Great Northern Railway* (GNR) where he worked for their Chief Mechanical Engineer, Henry Ivatt. Some of Ivatt's locomotives had problems and Gresley solved them. Because of this he quickly found himself made *Chief Mechanical Engineer* of the LNER.

Gresley's Worries

Ten years later in 1933, A3-Pacifics were looking out of date, though they were still among the fastest locomotives on British rails. The LNER wanted Gresley's team to do it again for a new, even faster, locomotive. This was to be the *A4-Pacific*. It had to be more than just a change of number. Where could Gresley and his teams start?

The name *Pacific* had been thought up by the GNR in 1923. They had given this name to the layout of these locomotives' wheels. This was '4-6-2'.

These numbers are called the 'Whyte notation'. The first number is the number of small wheels before the big driving wheels, the middle one is the number of driving wheels, and the last is the number of wheels after the driving wheels. Small ones at the front take the weight of the front of the locomotive and help to steer it on the tracks. Small ones at the back take the weight of the back of the locomotive.

The total number of wheels and the size of the driving wheels decide what sort of work the locomotive is good for. 4-6-2 is good for fast express locomotives. 0-6-0 is good for pulling slow, heavy freight trains. That's just six driving wheels with no small ones in front or behind them.

In the early days, 2-2-2 or 4-2-2 were used a lot for express locomotives. The single driving wheel each side was very big to get a high speed.

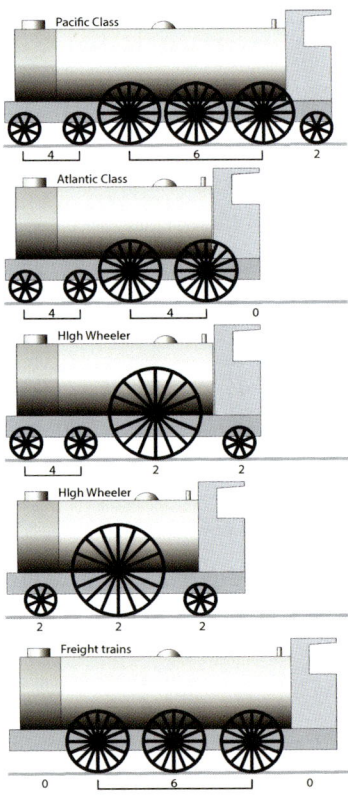

Some locomotive wheel layouts and their Whyte Notation. In the USA, the number of wheels on one side is used. (Image T. Farris.)

A New Concept

In the 1930s, people were talking about speed and something called *streamlining*. Aeroplanes, racing cars and road cars were all losing their clunky looks to become sleek and smooth with streamlining. Wasn't it about time railway trains did the same?

A 1920 Bugatti Type 13 Brescia racing car lifting a wheel on a corner before the days of streamlining, and a Bugatti Type 32_'Tank' with streamlining. This car led the way in streamlining.
(Photos by Kahzu on Pinterest, and by Arnaud 25 in the Public Domain.)

Let's face it, stunning as they look, steam locomotives were just downright clunky. Nothing about them was sleek. At low speeds this didn't matter. At high speeds, their blunt front caught the headwind. Then down the sides, bits and pieces sticking out caught the wind too. All this slowed locomotives down and made them use more coal and water. Even if that weren't too serious, people thought a locomotive that didn't at least look streamlined, was out of date. Something had to be done, and quickly.

The Flying Hamburger

Gresley heard about a very fast diesel-electric railcar in Germany. This was faster than any train in the world at 160 km/h (100 mph). It was called the 'Fliegende Hamburger'. What's that in English?

The German 'Hamburg Flyer', the fastest train in 1932. (Photo Bundesarchiv, Bild 102-14151 / CC-BY-SA 3.0.)

It's 'Hamburg Flyer' or 'Flying Hamburger' (not a bun with meat in it).

He noticed that its front wasn't flat but pointed a bit like the front of a boat. At high speed this would cut the air like a knife and make it stream smoothly down the sides and over the top. The air wouldn't then slow down the railcar so much. The next question is how did the engineers get the best shape?

Getting the Wind Up

They used a *wind tunnel*. Instead of trying to whizz a train along, or even a model of one, a model was held in place and a strong fan at one end blew thin smoke through the tunnel. The tunnel was big enough to have a small model of the railcar put in it. The model would be made of Plasticine so it could be easily moulded into different shapes.

When the fan was switched on, the engineers could see the smoke streaming around the model. If they saw smoke not flowing smoothly at some point, they would stop the fan. After opening a door in the side of the tunnel, they got inside to re-shape the Plasticine and try again. Bit by bit, they would get the best shape they could. The next stage was to take photographs and measure what they had done so that the shape could be made bigger on the real railcar. This was all new to Gresley and made him think.

How to streamline a steam locomotive that had a big blunt front? Searching for an answer he went to France to look at another railcar he'd heard about. This one was designed like a racing car.

National Physical Laboratory wind tunnel experts with model LNER locomotive, 1932 [Institution of Mechanical Engineers (PHO/NC/2]

Racing Ahead

The designer was the famous racing car designer, Ettore Bugatti. His family came from Italy but moved to France. Bugatti himself became French. He had a problem though. As well as making racing cars, his firm made cars for normal roads.

Bugatti's own very big luxury 'Royale' car of 1929. Cars like this carried over from the days of coach and horses with the driver outside in front and the passengers in luxury inside looking down their noses at people. A very long bonnet showed people you were rich enough to have a very big engine too, like having lots of horses. This car had a 12.7 litre, eight cylinder, engine. (Photo https://commons.wikimedia.org/wiki/User:Thesupermat.)

One of these cars had a very big 12.7 litre (775 cu in) eight-cylinder petrol engine under the bonnet. He only sold a few of these cars, then none at all. This left him with lots of very big engines lying about that couldn't be used in anything but the biggest cars that nobody wanted. How was he going to get rid of them?

He would have thought a lot and maybe talked it over with his engineers. One way or another they came up with a great idea. It meant a lot more work, but also gave his firm the chance to make more money. It wasn't another car. It wasn't a truck or a bus, or anything else for the roads.

It was a fast railcar, and it was pure genius. The genius was putting two Bugatti inventions together, one big, the other little. The big one was the enormous engine for his luxury car, of course. The other was the streamlining from his tiny Bugatti *Tank* racing car. Who else would have thought of it?

Bugatti's 12.7 litre, eight cylinder, petrol engine.
(Photo 'Petrolicious' via Pinterest.)

Together, these would make the railcar very fast and much easier and cheaper to keep running than steam trains. Then because it had two 12.7 litre engines, Bugatti could sell even more of those engines. He had a railcar built and tested, and showed it to the French government. Did they like it?

Bugatti's fast railcar of 1933. The driver sat in the pod on the roof. *(Photo and....T on Flickr.)*

You bet they did. They bought more than 100 of them. Some had two engines powering a single carriage or two carriages. Others had four engines for three carriages. One broke the world speed record at 196 km/h (122 mph). Everyone was happy. It solved a railway problem and Bugatti's problem with spare engines. His factory even had to build lots more. It was win, win. Gresley rushed back to London to tell his people.

The Thumb Print

So Bugatti's streamlining was the secret behind the A4-Pacifics. Gresley (or one of his team) had already worked out that his A3-Pacifics with more power, and with *streamlining* of the whole train, could pull eight or nine carriages at the same speeds as the small railcars Gresley had seen, and faster than the A3-Pacifics themselves. It was now a matter of getting down to work.

First, they used an A3-Pacific called *Papyrus* to test ways to get more power. They made the boiler pressure higher, made the steam pipework smoother, made the firebox bigger and doubled up the chimney. Did this make much difference?

It really did. Trains pulled by *Papyrus* now reached a record speed of 175 km/h (108 mph), whizzing from London to Newcastle upon Tyne in under four hours. This was a record. That was good enough. The bosses told Gresley to go ahead with the next stage.

A3-Pacific 'Papyrus' used for testing ideas for the A4-Pacifics.
(Photo Historical Railway Images, Creative Commons Attribution-Share Alike 4.0 International license.)

It's likely Gresley's team wouldn't know much about streamlining, and a wind tunnel was needed to help them do it. One of the office ladies booked time in the wind tunnel at the National Physical Laboratory in London. Gresley asked a Professor Dalby to help come up with a streamlined shape based on Bugatti's racing car and railcar. As Bugatti's engineers had done before, Dalby made a Plasticine model. The engineers ran test after test, with small changes to the shape of the front. They were getting there but for one problem that they just couldn't solve.

This was a problem that a lot of steam locomotives had. It was smoke from the chimney blowing around and stopping the driver and fireman seeing ahead. The engineers had the same problem in that wind tunnel. They kept making changes to the shape. Nothing they did solved the problem. They even thought of fitting big, ugly 'smoke deflectors' as many locomotives already had. They knew this was pointless because they would ruin the streamlining. What to do? It was the end of the day. They were stuck with a problem they couldn't cure. They were tired and fed up. "Oh, just one more test, then we'll go home and hope to solve it tomorrow."

Miracle! The smoke flowed over the top of the locomotive. It was just what they wanted. How did it happen?

Smoke deflectors on the Flying Scotsman. These show how worried the engineers were in thinking about these for the A4-Pacifics. (Cropped from photo by Rich@rd in the Public Domain.)

They stopped the fan, opened the wind tunnel and looked carefully around the model. What was different? Nothing. Then one of them noticed something. "Hey, look at this! There's a thumb print in the Plasticine just behind the chimney. It could be that."

So they smoothed it out and ran another check. The smoke went all over the place as before. The engineer with the clumsy thumb made a new thumb print. They ran another test. The smoke went over the top again. The roughness of the thumb print was ruffling the wind behind the chimney, forcing the airflow upward, taking the smoke with it. Success! The engineers went home happy, wondering how to add a big thumb print on the real locomotives.

Of course, the real locomotives weren't made with a big thumb print behind the chimney. Instead, the designers came up with something else to ruffle the wind. So a big part of the success of the

A4-Pacifics was because of a brilliant idea by a great Italian designer and a thumb print by a tired British engineer.

Mallard, the most famous A4-Pacific.
(Photo adapted from PTG Dudva Creative Commons Attribution-Share Alike 3.0 Unported.)

What the Papers Said

Designing the A4-Pacifics was more than just coming up with a new locomotive. They were to pull the new *Silver Jubilee* service between London and Newcastle on Tyne. The whole train had to look the part. The designers worked their way back along it to make everything else streamlined. This meant the *Silver Jubilee* trains with their silver A4-Pacifics had their own matching carriages. That's not all the engineers worked on though.

Inside, they were the most comfy of British trains at the time, and they rode smoothly on the tracks. These had to be kept in perfect condition to help give that good ride and to be safe at high speeds. It was a whole *package.*

This was of comfort, safety and speed, not just a locomotive. The LNER were very proud of their new train, telling the newspapers what they had done. In typical British fashion, the newspapers found fault as much as they could.

Some newspapers, who thought they knew better, said the streamlining was just *hype*. Others that the LNER were playing at following the latest craze, and they didn't really know what they were doing. This must have upset the people who had worked so hard on designing and building the train. The newspapers were wrong, of course.

They looked foolish when different A4-Pacific locomotives broke one speed record after another. They could also run non-stop further than any other locomotive. This was the nearly 640 km (400 miles) between London and Edinburgh. It was quite a feat that amazed people in much of the world.

Streaking

The A4's could do it because they could make a tender-full of coal last the whole way. Also they allowed crews to change over on the move with a corridor in the tender. This led them to and from the first carriage to rest or go to work on the footplate. Streaking along non-stop as they did, people started calling them the 'Streak'.

Silver and Gold

The first four A4-Pacifics had the word 'Silver' in their name because they were used on the *Silver Jubilee* service. These four kept their names all the time, but others had their names changed at some point. Three had the word 'Golden' in them. Five were named after countries in the British Commonwealth. Seven were named after famous men and bosses of the LNER including Gresley. Most of the rest were named after birds. Among them was one bird that was rather special because it was the bird Gresley was fond of. He even bred them himself.

It was a type of duck, one you hear quacking on tiny village ponds. It was, of course, *Mallard*. As if changing names wasn't enough, the LNER changed something else too.

That was colour. Every two or three months, express locomotives had to be taken apart, cleaned inside and out, and put back together with new parts. By this time the paint work could be looking shabby, so it may be freshened up. Rather than just repainting them with the same colour, the LNER sometimes played around with different colours. Maybe it was to make people think the locomotive pulling a train into a station would be an exciting new one, not an old one with a new name and fancy colour scheme.

The different colours of the A4-Pacifics. (Image Ace Models.)

Pushing the Limits

Breaking speed records wasn't easy. Even normally, driving express steam locomotives needed special skills. Driving one as fast as it could go needed more of a skill, being *at one* with the locomotive, and being daring.

Some drivers became well known in the LNER for being like this. One of them was George Haygreen with his fireman Charlie Fisher. When the LNER wanted to have a go at breaking the speed record in 1935 they got him to drive 'his' fairly new A4-Pacific *Silver Link*. The journey was just over 180 km (110 miles) from London to Grantham. During it, he got the train up to a record 182 km/h (112.5 mph). Time for the LNER to celebrate!

It was all rather silly really, just for a railway company to say that one of its locomotives was just a tiny bit faster than a rival's on that day. It was also risky because they were using tracks that weren't rated for those speeds, and the trains were carrying passengers too. It was just as well that no accidents happened on the LNER because at such high speeds, trains become totally wrecked and people get killed. (This happened on the LMS.) Being in a train wreck wasn't what they bought their tickets for. It didn't stop the railway firms from trying again and again with a train full of passengers though.

A year later in 1936, the LNER got George Haygreen to drive their newest A4-Pacific, *Silver Fox* with Charlie Fisher. The fireman is just as important as the driver because he has to make sure the steam pressure and other things are right for what the locomotive has to be doing next. If it is to start climbing or the driver is told to drive the locomotive hard to try a speed record, the fireman must raise the steam pressure well ahead while the driver starts to speed up the locomotive.

The LNER had a favourite run of track for trying to break a speed record. This is called *Stoke Bank*. Going south it slopes gently downhill between Grantham and Peterborough, helping trains build up speed. It does seem cheating, but nobody seemed to mind. For a record-breaking run, or sometimes on a normal run, a railway inspector would ride on the footplate.

Driver George Haygreen with his fireman Charlie Fisher. Fisher's right hand is black from the coal, and he's pressing his left hand into his back to ease the pain of shovelling coal for hours. (Photo LNER Encyclopedia.)

While trying to keep out of the way of the busy fireman, he would be taking readings from the gauges and working out how much coal and water has been used. Afterwards, he would give these to the engineering people for them to see how their locomotive was working.

On this journey, an inspector was on the footplate but the crew thought nothing of it. Haygreen drove the locomotive normally. Just as the train got to the top of Stoke Bank, the inspector suddenly told him to try for a speed record. He was horrified.

The train wasn't going fast enough, to start trying to break a record. The steam pressure was too low. Fisher swung round. Bending double, he thrust his shovel under the mound of coal in the tender behind him. He swung back and tipped the coal into the fire, not any old where, but where the layer of coal was looking thin on the grate. Back and to he went, shovelling coal like fury onto the fire, but not too fast for fear of sudden extra heat cracking the firebox walls.

The locomotive started to use water more quickly. He kept checking the water-level gauge, letting more water in, but not too fast for fear of cracking boiler tubes with the sudden change in water temperature. Is the steam pressure going up?

This mustn't happen too fast for fear also of cracking boiler tubes. Sweat was pouring down his face as he shovelled and checked, shovelled and checked, on and on, giving Haygreen as much as he dare. All the while he was worrying about damaging the firebox and boiler.

While keeping an eye on the pressure gauge, Haygreen nudged the speed regulator further and further. From time to time he span a hand wheel in front of him to change something called the *offset*. He was squeezing as much power as he could from Fisher's rising steam pressure. The speed galloped up. There wasn't much time left. Not far from the end of the bank he would have to slow the train down a lot. Will they get up to speed and slow down in time?

hard as he dare without throwing the passengers about. Even though he was driving about 80 tons of locomotive, he had to be as gentle as if it were a baby. The speed trial was over almost as soon as it had begun. Haygreen and Fisher couldn't have done more. Did they break the speed record?

Top speed was 183 km/h (113 mph), beating their previous record by just half a mile an hour. Since they started the speed test so badly, that was a credit to their efforts. The locomotive paid a price for it though. Working so hard had overheated a problem bearing and caused it to start melting. Haygreen would have felt the locomotive starting to run rough.

A tired driver at the controls of an A4-Pacific.
(Photo A. Wills.)

The train was thundering along the track. At the last moment, Haygreen shut down the speed regulator and braked as

He could tell what it was. He slowed the locomotive down even more to save the bearing. Would it get worse and make the locomotive break down?

Nursing the locomotive the rest of the way caused the train to arrive in London seven minutes late, but it did get there. It was a hopeless speed test, of course, but it didn't stop the LNER from still trying to break speed records. They just needed a way to look after that bearing and get a tiny bit more power. Where could they get it from?

They had to wait two years to find out. In the meantime, the LNER's rival, the LMS, on the West Coast Main Line beat the LNER's record with their new streamlined *Coronation Class* locomotive. Its top speed was 185 km/h (114 mph). Again just a half a mile an hour more. Time for the LMS to celebrate! Not for long though. The LNER were working on something new.

Rival Coronation Class locomotive of the LMS that broke the A4-Pacific record.
(Photo posted on Reddit by i.imgur.com/GkPVot.)

The Train that Thought It Was a Racing Car

It was the summer of 1938, a year that railway lovers never forget. Gresley was worried that brakes needed to be better now that trains were running much faster. He had a new braking system fitted to a train and set up trials. Actually, the braking system wasn't that new. It was the *Westinghouse* system that their rivals the LMS had been using for quite some time. The way they wanted to test the brakes was to run up to high speed as quickly as they could, slam on the brakes and see what happens. All they needed was a really powerful locomotive to do it.

It so happened that *Mallard* had just been built. It had something the other A4-Pacifics didn't have. That was a *Kylchap Double Blast Pipe*. This gave it more power. There was one other thing the locomotive needed. That was a crew of driver and fireman who knew how to drive locomotives hard. This time the LNER chose Joseph Duddington with fireman Thomas. Bray. (Probably Haygreen and Fisher had had enough of speed trials and wanted a quiet life.) Also on the footplate in smarter clothes, giving orders and taking measurements was an inspector, Sid. Jenkins. Everyone was sworn to secrecy.

This was for two reasons. The LNER didn't want the rival LMS to hear that the LNER were to do a speed trial on the way back. Also the engineers didn't want the people who look after the track to know they would be running at 190 km/h (120 mph) or so on tracks rated for 145 km/h (90 mph). There was one problem that was facing them though.

It was the bearing that overheats and maybe breaks up when an A4-Pacific is driven hard. After *Silver Link* had been put out of action for a few days two years before, they didn't want to do the same to *Mallard*. It was a tense time.

They ran the brake trials from London to Peterborough about 138 km (85 miles) away. Before setting off on the return trip, Duddington poured as much oil as he could over the bearing and strapped a *stink bomb* onto it to warn him if the bearing was overheating. The bomb would burst and a strong smell of aniseed would fly out. Duddington crossed his fingers. Would it hold? It was wait and see.

When they got near to the top of Stoke Bank on the way back, the driver and fireman were prepared to go all out for a record-breaking run. The train gathered speed: 39 km/h (24 mph), 121 km/h (74.5 mph), 152 km/h (94 mph), 188 km/h (116 mph; speed record broken!), 194 km/h (120 mph), then 202.5 km/h (125 mph). That was it. This speed lasted for just 278 metres (306 yards) before the train had to be slowed down.

Where is Stoke Bank?

If the LNER hadn't used Stoke Bank for squeezing as much speed as they could out of their steam locomotives, it's likely hardly anybody would have heard of it.

Stoke Bank shown in red. (Image Google Maps.)

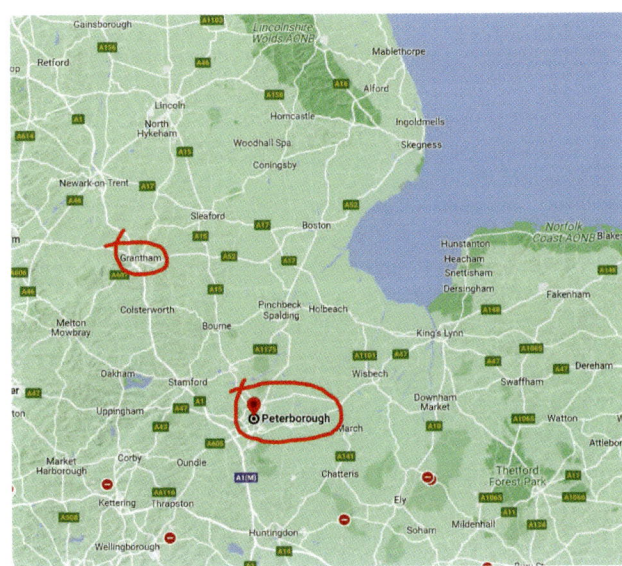

Where Grantham and Peterborough are on the east side of England. (Image Google maps.)

It's just a straight bit of railway track that slopes gently downhill going south between the villages of Little Bytham and Essendine stations. These are between Grantham and Peterborough.

Afterwards, Gresley said that 125 mph was *Mallard's* top speed on that run. Some people said it was 126 mph. The reason for this is that a blip on the paper roll that recorded speed showed 126 mph for a very short distance. So take your pick. Was it 125 mph or 126 mph?

What Mallard looked like under the streamlining.
(Photo source unknown.)

In the end, *Mallard* pulling a good size train was a tiny bit faster than the tiny German *Hamburg Flyer*, and the LMS's speed record. Surely that was close enough in itself without fussing over one half a mile an hour? We're forgetting that problem bearing though.

It didn't hold. It was bad. It had melted during the speed test. Smelling aniseed, Duddington had slowed the train down even more afterwards. He stopped it at Peterborough rather than carry on with it to London. Another locomotive took the train the rest of the way. The LNER were prepared for this, handing out photos of *Mallard* to the newspaper reporters at the London station. They also gave the reporters a made-up story why an older locomotive was pulling the train in.

On another run of a Pacific locomotive, the bearing had overheated so much that it had broken up. This caused the piston that drove it to crash into the ends of its cylinder and smash it to pieces. Bits of metal would have been flying everywhere. That was bad. In normal use though, running at up to 160 km/h (100 mph) and when properly looked after, the bearing gave no problems.

It's nice to say *Mallard* has done 125 mph or 126 mph but it was only downhill for a few seconds and it damaged itself doing it. A4-Pacifics could never have been racing cars.

Oh No! Not Another Speed Trial

By 1948, all the railways were being run by *British Rail*. At the time, the country was busy rebuilding itself after the Second World War. Never mind that, *British Rail* decided to hold speed trials between the express locomotives from their four railway companies.

They were the LNER, LMS, Great Western, and Southern Region. As for the A4-Pacific we know what would let it down, and it did. The happier story was that it was the most frugal of them all on coal and water. Did BR learn anything from the trials?

Who knows!

That Bearing Problem

When the A1-Pacific was being designed by Gresley and his team, he had decided to give it three cylinders. One was on each side and one in the middle underneath the smoke box. This is at the front of the locomotive. Like a three-cylinder car engine, three cylinder steam engines ran smoother than two or four cylinder ones. There was a downside though.

This was how to make sure steam was fed to the middle cylinder at the right moments. Gresley and another engineer named Harold Holcroft had worked out a way to do this. They came up with the *Gresley-Holcroft conjugated valve gear*. Never mind the name, it was a forest of links swinging back and to underneath at the front. At high speed, these links failed to keep the steam timing correct. The middle cylinder ended up working harder than it was supposed to. This over-worked the bearing it was acting on. This is what caused it to overheat and even break up. Was there a cure?

Later on, the engineers changed the bearing for a bigger one, but that didn't deal with the cause of the problem. This was fixed, but not until around 30 years later in the 1950's when speed trials were a thing of the past. It's a pity it couldn't have been done earlier, since Haygreen felt that the A4-Pacifics could have got up to 219 km/h (135 mph) if it weren't for this problem. That would have showed them.

Was It All Worth It?

Driven as they should, the A4-Pacifics gave great service, especially on the *Silver Jubilee* run to and from Newcastle Upon Tyne, running at up to 160 km/h (100 mph). More and more passengers took the train. There were even more of them than the LNER had hoped. They had to add an extra carriage to carry them all. It wasn't a long train though, just eight carriages. Everything was going fine, then a year later, it all went wrong.

World War Two started. Services were cut, trains had to run slower to save coal, and freight took priority over passengers.

To begin with, the *Silver Jubilee* trains that were the pride of the LNER were put in storage. Other A4-Pacifics were used for pulling heavy freight trains, something they weren't designed to do. A streamlined A4 pulling a train of coal trucks was a sad sight. It was enough to make people weep. The A4's did this alright except they broke down a lot because of the middle bearing problem being made worse by poor upkeep. Instead of scheduled upkeep, it became continuous crisis. Then the final humiliation came.

The A4's had their own style of whistle which was called a *chime whistle*. The government told the LNER to remove and destroy these in case people hearing one thought it was an air-raid warning siren, would you believe! Standard whistles were fitted instead. As soon as the war ended, the LNER fitted new chime whistles. Worse, while *Sir Ralph Wedgwood* was parked at York, a wartime bomb blast so badly damaged it, it had to be scrapped.

Sir Ralph Wedgwood outside York North Depot after a bomb blast in World War Two.
(Picture reproduced courtesy of The Press, York......www.thepress.co.uk)

Let's Get Back To Normal!

It never fully happened. Just as the LNER were about to get back to non-stop services, several railway bridges in Scotland had collapsed due to flooding. Trains had to go to the other side of the country, which added nearly 16 km (10 miles) to the nearly 640 km (400 mile) journey. A new record was made, though.

The LNER kept the train going non-stop, making it the longest non-stop run ever. The LNER was happy. This happiness wasn't to last long though.

Trains ran slower than before the war. Non-stop services soon stopped running anyway. Prestige and pride were replaced with just doing a job. Staff didn't really care any more, Stations and trains became grubby and scruffy. The great days were over.

Steam locomotives were getting out of date. They were dirty and cost a lot to keep running, Diesel engines and electric motors were now more powerful and reliable than they used to be. Steam locomotives were quickly scrapped. Diesel and electric power took their place. On the 14th September 1966, the last A4 ran just between Aberdeen and Glasgow. There was no glory, no cheering crowds, just a quiet goodbye. Gresley's A4-Pacifics deserved better.

Cherished

Perhaps British Rail remembered the painful lesson of the A3-Pacifics which were all mercilessly scrapped except for one, the *Flying Scotsman*, and that was only because of the name. Rich people, one of them after the other, had stepped in to save the locomotive from the cutter's torch. Two of them lost everything trying to pay off debts caused by it. The first one was saved. The second died young after losing everything he owned, and nobody rose to save him. This couldn't be allowed to happen again.

Five A4's were kept. Three are in Britain, one in North America, the other in Canada. Keeping steam locomotives in running order costs a fortune, as people who had tried to save the *Flying Scotsman* had found. It is a credit to the railways with four of the A4's that they are keeping theirs running.

As for the fifth one, *Mallard*, it last ran in 1988 on the 50th anniversary of its record-breaking run down Stoke Bank. Since then it can no longer be put in steam. It is being kept on display at the National Railway Museum in York.

Thank goodness for that.

On the next page: Mallard.
(Photo adapted from Mallard_in_York - PTG Dudva Creative Commons Attribution-Share Alike 3.0 Unported.)

For more fascinating *Great Steam Train* stories from Tom Farris, see…

Britain's First Railways
Flying Scotsman
Trains Racing North
Princess Royal and Coronation Scot
The Grand Dream of Broad Gauge

Hamilton-Vale books are available via all good bookshops or online from Gwales.com, Amazon.co.uk and other website sellers.

Be the first to know about our new titles. Join our mailing list. Send an email with '**JOIN**' in the subject to info@graham-lawler.com